Rocks

© 2015 OnBoard Academics, Inc
Portsmouth, NH
800-596-3175
www.onboardacademics.com
ISBN: 978-1-63096-048-3

OnBoard Academic's books are specifically designed to be used as printed workbooks or as on-screen instruction. Each page offers focused exercises and students quickly master topics with enough proficiency to move on to the next level.

OnBoard Academic's lessons are used in over 25,000 classrooms to rave reviews. Our lessons are aligned to the most recent governmental standards and are updated from time to time as standards change. Correlation documents are located on our website. Our lessons are created, edited and evaluated by educators to ensure top quality and real life success.

Interactive lessons for digital whiteboards, mobile devices, and PCs are available at www.onboardacademics.com. These interactive lessons make great additions to our books.

You can always reach us at customerservice@onboardacademics.com.

Rocks and the Rock Cycle

www.onboardacademics.com

igneous	sedimentary	metamorphic
Cooled lava or magma.	Sand and other materials collect on the ocean floor and are compacted together.	Rock deep in the earth, under heavy pressure and heat forms new minerals and textures.

A rock is a solid mixture of crystals of one or more minerals. Rocks can also be formed from organic material. Rocks fall into three main categories based on how they are formed: igneous, sedimentary, and metamorphic.

Igneous Rock

Igneous rocks are formed when molten rock called magma rises near or onto the Earth's surface and cools.

We call igneous rock that forms below the Earth's surface intrusive igneous rock and igneous rock that forms on the Earth's surface extrusive igneous rock.

When magma cools slowly underground to form intrusive igneous rock the individual mineral grains have a very long time to grow. Intrusive igneous rock therefor have large crystals and a course grain texture.

The most common intrusive igneous rock is granite. Diorite is another example of an intrusive rock

Intrusive Igneous Rock

Extrusive Igneous Rock

Extrusive igneous rocks are formed when volcanos erupt and magma is forced to the earth's surface. We call magma that reaches the earths surface lava. When lava is on the earth's surface it cools quickly so the mineral grains have little time to grow which is why extrusive igneous rocks have a fine grain or even glassy look.

Gas bubbles are trapped within the rocks giving them a sometimes bubbly texture.

Basalt and pumice are examples of extrusive igneous rock.

Sedimentary Rock

Sedimentary rock is formed when rock from mountains and the earth surface are broken down as a result of erosion and weathering. These small pieces of rock along with other materials such as clay and silt are carried away by rivers and streams to oceans and lakes.

This material which we call sediment eventually settles in layers in oceans and lakes. Over time the pressure from the water and the chemicals from the water cement the material together to form sedimentary rock.

Often time animals and plants are trapped in the rock and form fossils. Scientists can form a time line based on the the sequence of layers and fossils found in sedimentary rock.

Metamorphic Rock

Metamorphic means a change in form and metamorphic rock is formed when igneous and sedimentary rock are forced deep within the earth. Heat and immense pressure squeeze the mineral grains and sometimes rearrange the atoms to form new minerals. The heat and immense pressure creates a new type of rock that we call metamorphic rock.

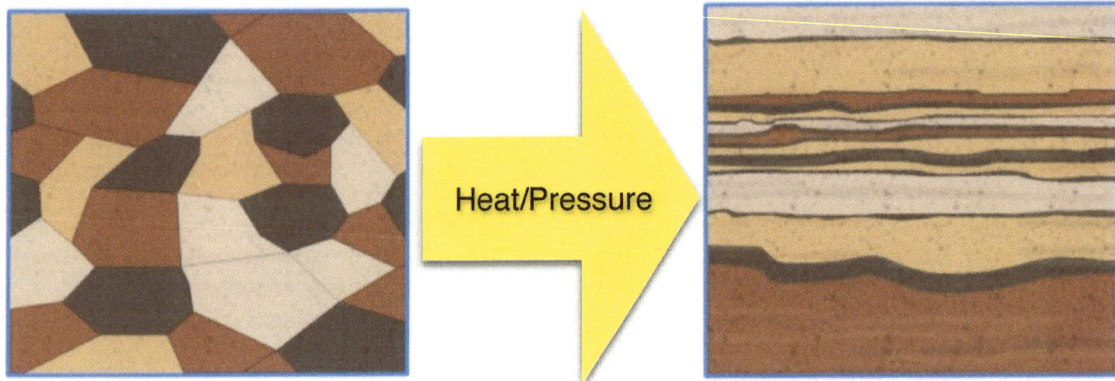

Heat/Pressure

Summary

Igneous rocks are formed when molten rock cools. We call igneous rock that forms below Earth's surface intrusive igneous rock, and igneous rock that forms above Earth's surface, extrusive igneous rock.

Sedimentary rock is formed when small pieces of rock and other materials settle in layers at the bottom of oceans and lakes. As more and more layers accumulate, the sediment in the bottom layer is cemented together to form sedimentary rock.

Metamorphic rock is igneous or sedimentary rock that is forced deep into the Earth. Immense heat and pressure form a new rock that we call metamorphic rock.

Rocks

Match the vocabulary word with the correct definition.

magma	This type of rock forms in layers and often contains fossils.
metamorphic	This type of rock is formed from cooling magma or lava.
extrusive	This type of rock is formed deep within the earth under intense heat and pressure.
igneous	This describes igneous rock that forms at or above the Earth's surface
intrusive	This describes igneous rock that is formed below the Earth's surface
sedimentary	This is the name we give to molten or liquid rock.

www.onboardacademics.com

The Rock Cycle

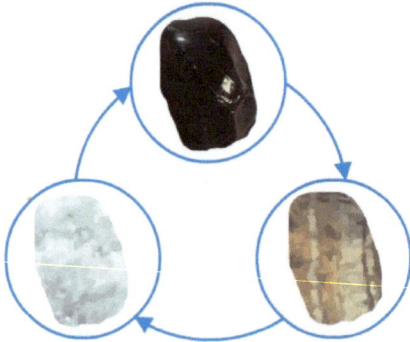

The rock cycle describes the continuous process in which rocks cycle between each of the three types of rocks; igneous, sedimentary and metamorphic.

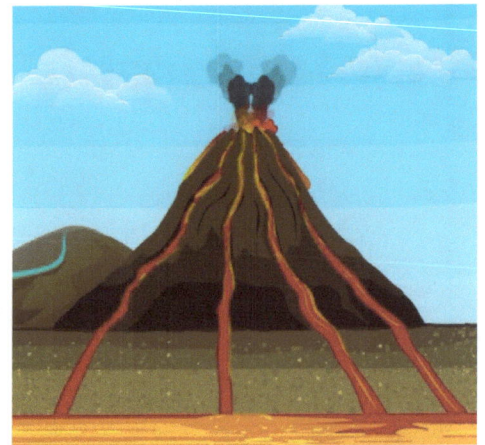

As we have learned, igneous rock is formed when magma cools at or below the Earth's surface.

Particles of the igneous rock are deposited on the Earth's surface as a result of erosion and weathering. These particles are carried with other materials to oceans and lakes.

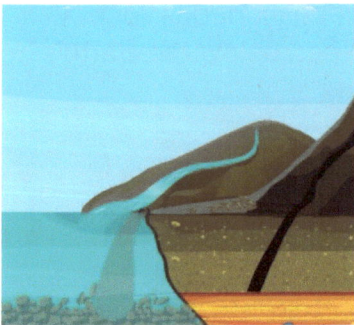

This material, that we call sediment settles at the bottom of oceans and lakes in layers. The bottom level of the layers is compacted and forms sedimentary rock.

Eventually the sedimentary rock is forced deep within the earth by plate movement where it is subjected to extreme heat and pressure. This transforms the sedimentary rock into metamorphic rock. This process can also occur with igneous rock.

Eventually the heat will turn the metamorphic rock into liquid magma. One day the magma that was once our metamorphic rock will return to the surface and start the cycle again.

The Rock Cycle

Number these steps so they are in the proper order representing the rock cycle.
Number one has been completed for you.

1 magma rises to surface

◯ sediment settles in layers at bottom of ocean

◯ metamorphic rock melts and turns into magma

◯ lava cools to form igneous rock

◯ magma rises to the surface

◯ sedimentary rock forced below surface

◯ particles and other materials carried to ocean

◯ bottom layer compacts to form sedimentary rock

◯ heat and pressure transform sedimentary rock

◯ erosion and weathering of igneous rock occurs

Name: _____

Rock Cycle Quiz

1. _____ rocks are formed when magma cools.

2. _____ rocks are formed when rocks buried under the Earth are compacted at high temperature and pressure.

3. _____ rocks are formed when grains of sand and other materials accumulate in layers on the ocean floor and are compacted and cemented together.

4. Igneous rocks formed above the Earth's surface are called _____.
   a. intrusive igneous rocks
   b. extrusive igneous rocks

5. Once upon a time, all metamorphic rocks were either igneous or sedimentary rocks. True or false?

6. Molten or liquid rock in the Earth is called _____.

7. The rising of magma to the surface is the first and last step of the rock cycle. True or false?

     www.onboardacademics.com

Minerals

www.onboardacademics.com

What is a rock made of?

Stone   Iron   Water   Steel   Minerals

Minerals are the building blocks of rocks which are formed from two or more minerals.

www.onboardacademics.com

What is a mineral.

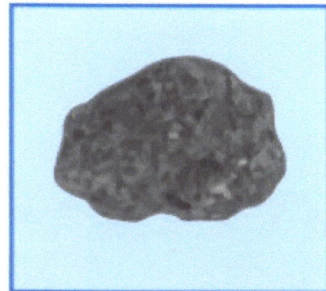

Minerals are inorganic (non-living) substances which are made from one or more chemical elements which form in a *crystalline* structure. This means that the tiny particles that make up the mineral (called atoms) are arranged in a very regular pattern such as a square or a double triangle. There are more than 3,000 different types of minerals on Earth, but only a few dozen are commonly found.

How many types of minerals are found on earth? _____

Do minerals contain living matter? _____

How many minerals are commonly found? _____

How many elements make up a mineral? _____

Where do minerals come from?

Most of our minerals are formed when molten rock, deep within the earth, moves toward the surface and erupts from volcanoes and then cools.

Elements inside the cooling magma come together in organized and repeating patterns to form minerals. A crystal is the name given to a solid block of minerals in this repeating pattern formation. If there is ample space minerals may form very large crystals. Talc is one of the softest minerals and diamonds are one of hardest.

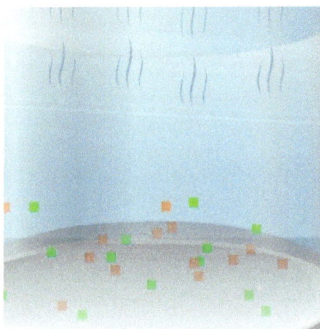

Minerals are also made when water that contains certain chemicals evaporates. For example, if you let salt water evaporate your are left with sodium chloride mineral more commonly known as salt.

Minerals are made when magma (liquid rock) cools into solid rock and also when water evaporates and leaves behind tiny pieces of minerals that were dissolved in the water.

Five characteristics of minerals

Hardness measures what things the mineral can scratch and what can scratch the mineral.

Color describes the mineral's color after you have removed dirt and other materials.

Cleavage describes what the mineral looks like when you break it.

Streak describes what kind of mark the mineral leaves when you try to write with it.

Luster describes if the mineral is shiny like metal or dull.

www.onboardacademics.com

On a scale of 1-10 how hard is each mineral? Use the hints below.

10	
9	
8	
7	
steel file → **6**	
glass → **5**	
4	
penny → **3**	
fingernail → **2**	
1	

- **Orthoclase** can scratch glass, but apatite cannot.
- **Quartz, Topaz, Corundum** and **Diamond** are all harder than steel.
- You can scratch **calcite** with a penny.
- You can scratch **gypsum** and **talc** with your fingernail.

www.onboardacademics.com

Are you a mineral master?
Fill in the blanks below.

All rocks are made of at least two different types of
_____. Minerals are made when _____ in
the Earth cools off, or when _____ evaporates.
There are five common properties of minerals: hardness, color, cleavage, streak, and _____.
_____ is one of the softest minerals; _____
is the hardest. The same mineral often appears in different colors, so to identify a mineral, you can use
a _____ test which means to rub the mineral on a
tile to see what color it makes. Cleavage is used to de-
scribe what a mineral looks when it is _____.
Luster describes if a mineral is shiny or _____.

streak   water   Talc   dull

magma   diamond   minerals   luster   broken

Minerals Quiz

1. _____ are the building blocks in rocks.

2. Minerals are inorganic substances made from a single element. True or false?

3. Minerals are crystalline in nature. True or false?

4. Minerals are made when _____ cools into solid rock.

5. Which of the following is one of the softest minerals?
    a. Quartz
    b. Topaz
    c. Talc

Plate Tectonics

www.onboardacademics.com

Pangaea

Scientists believe that many, many millennia ago the Earth had only one continent surrounded by one ocean. This continent is called Pangaea. See if you can recognize and label the current continents as they formed to make up Pangaea.

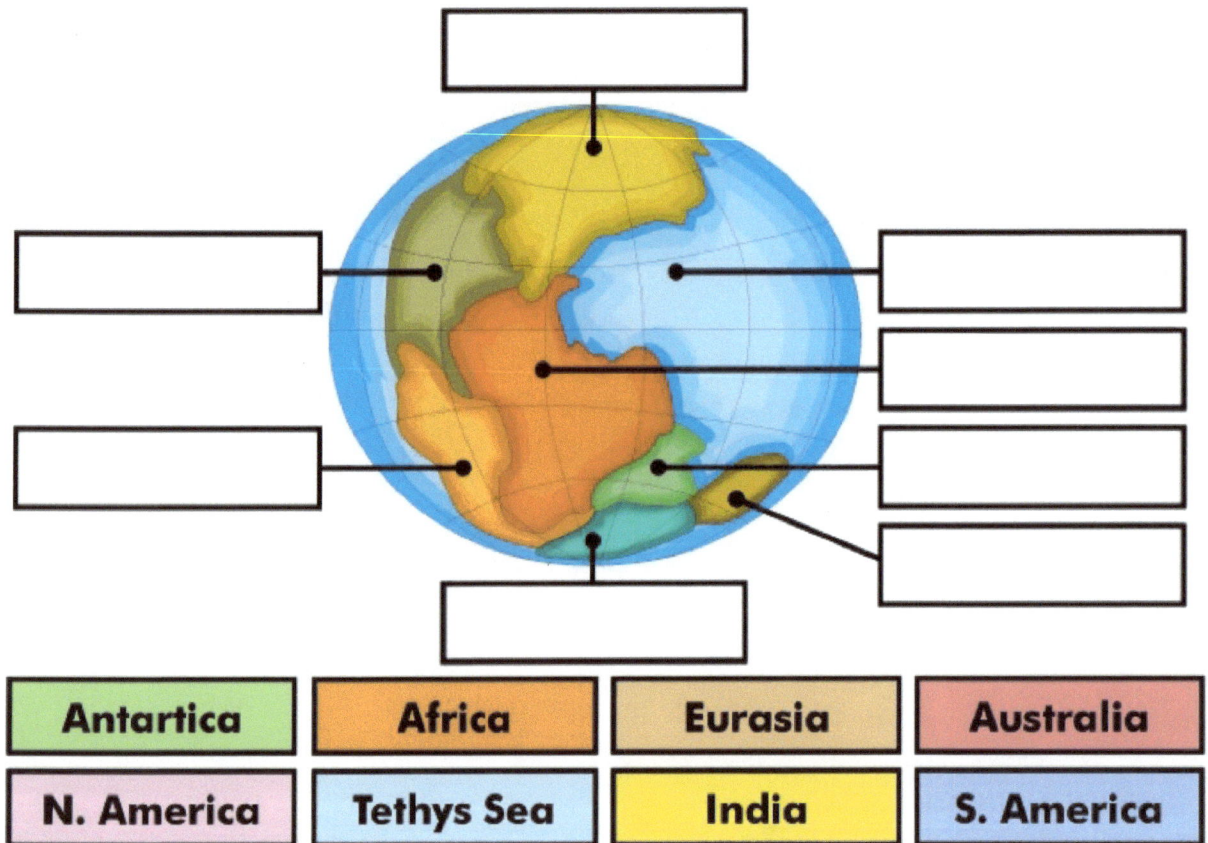

Antartica	Africa	Eurasia	Australia
N. America	Tethys Sea	India	S. America

A single ocean surrounded the super-continent Pangaea: the Tethys Sea.

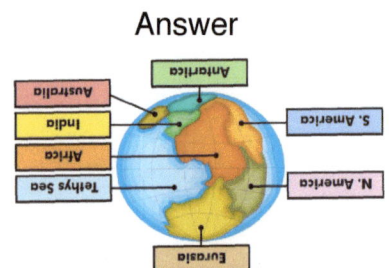

Answer

How were today's continents formed?

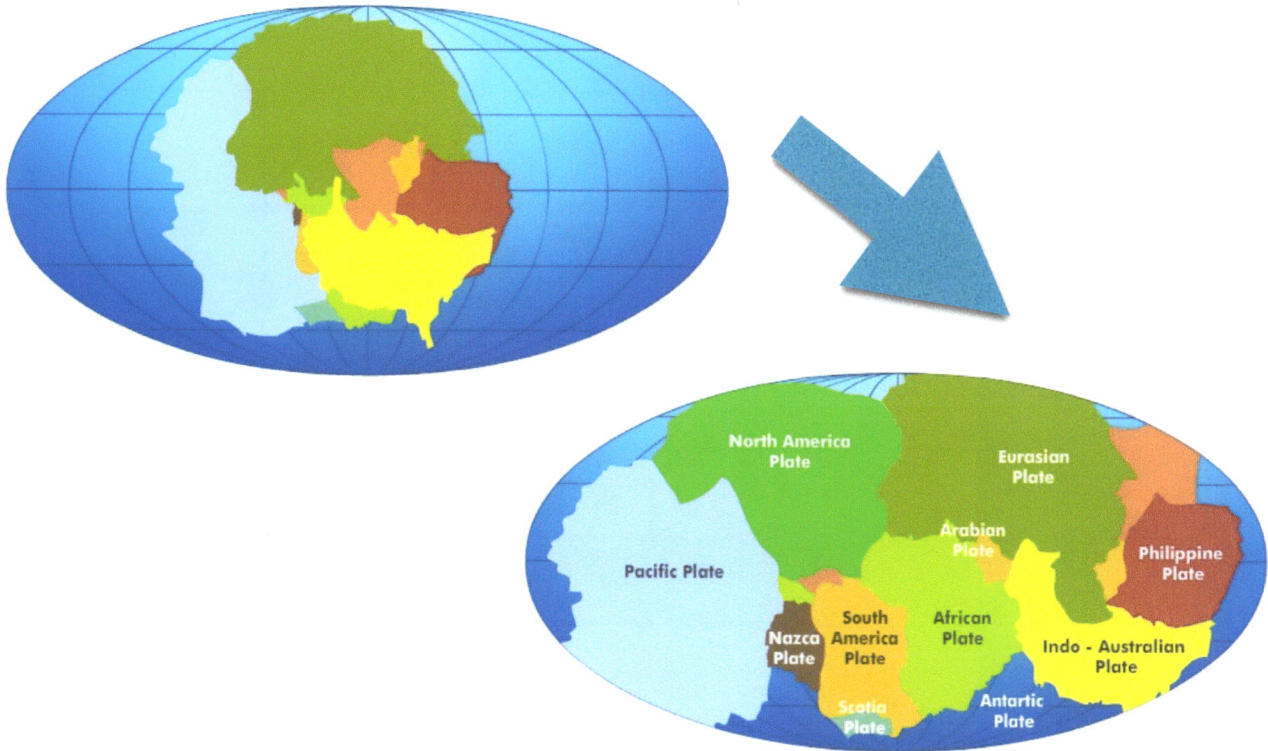

The Earth's crust floats on top of a liquid layer of rock called the mantle. The continuous movement of the mantle has cracked the Earth's crust into a number of large slabs called tectonic plates. As the mantle continues to churn and move, these tectonic plates, which were once linked together in a single continent, have slowly drifted apart. This is know as the continental drift theory.

Identify the seven major tectonic plates.

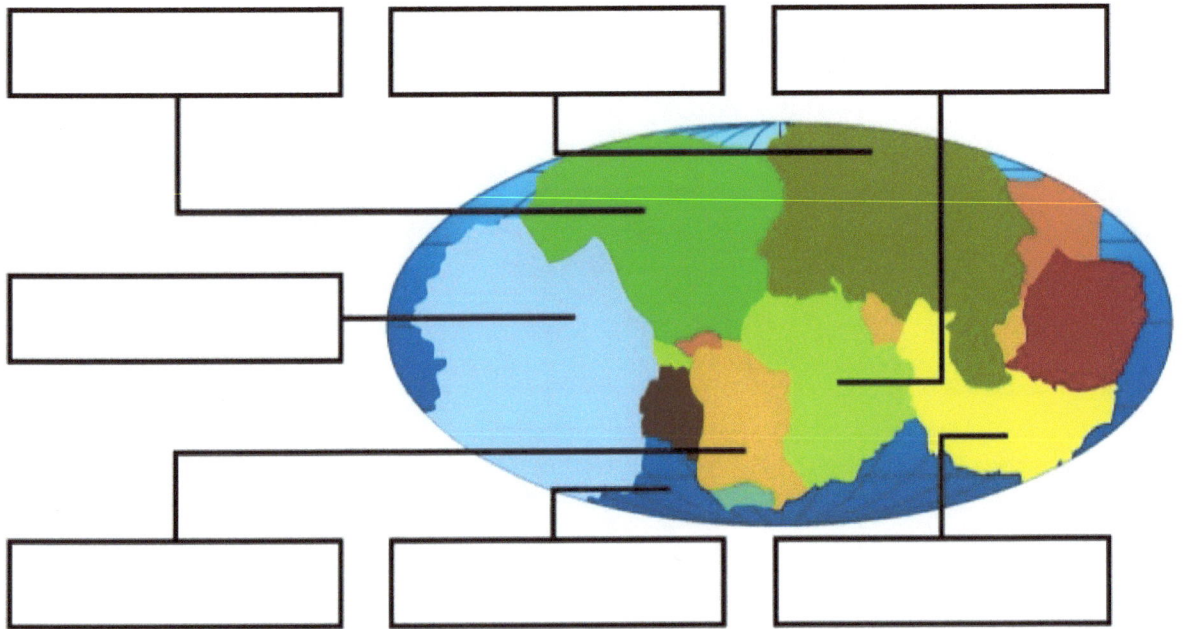

Eurasian	North American	African
	Pacific	
Indo-Australian	Antarctic	South American

www.onboardacademics.com

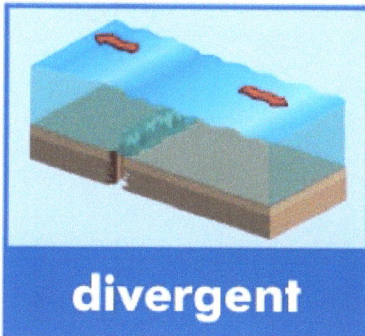

divergent   **convergent**   **transform**

There are three types of plate boundaries: divergent, convergent, and transform.

A **divergent boundary** occurs when two tectonic plates move away from each other creating a deep valley or "rift". Magma escapes into the rift and cools to form undersea mountains and volcanoes known as submarine volcanoes. Typically occurring under the oceans, divergent boundaries also have the effect of widening oceans by expanding the ocean floor.

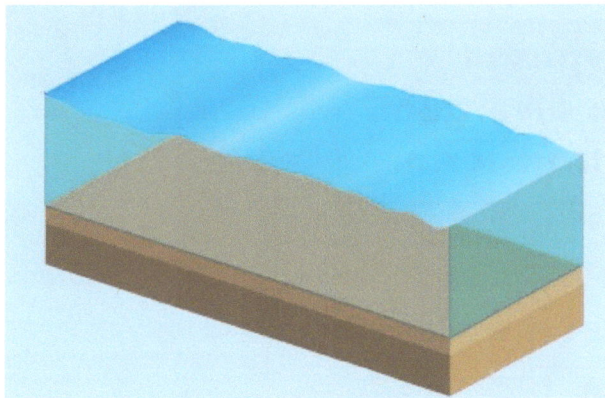

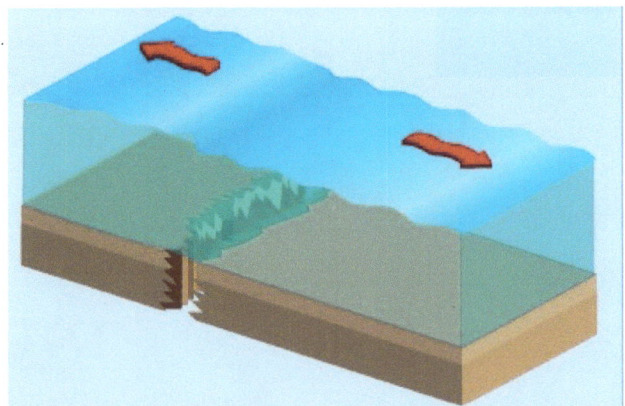

Convergent Plate Boundaries

Convergent boundaries occur when two plates slowly collide. When two continental plates collide, the crust folds to form mountains. When an oceanic plate collides with a continental plate, the oceanic plate is forced beneath the continental plate because it is thinner. This creates an oceanic trench known as a **subduction zone**.

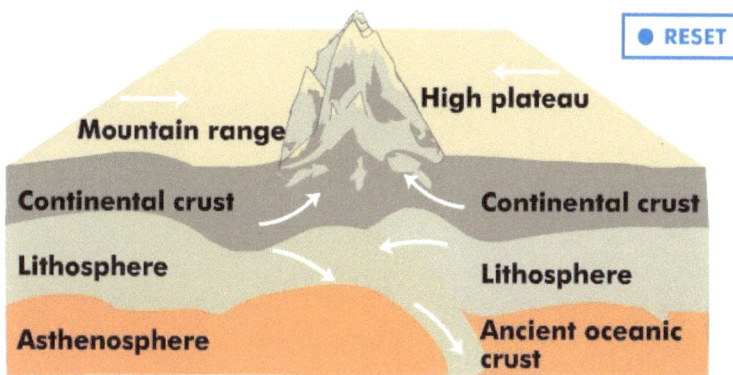

● RESET

High plateau

Mountain range

Continental crust

Continental crust

Lithosphere

Lithosphere

Asthenosphere

Ancient oceanic crust

Massive plates sliding horizontally past one another create **transform boundaries,** commonly known as faults. Transform boundaries can release an enormous amount of energy that can cause huge and very destructive earthquakes. A well-known example of a transform boundary is the San Andreas fault in California.

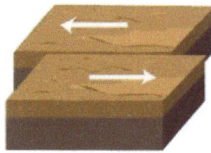

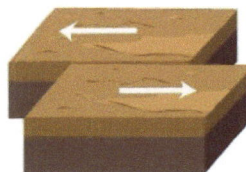

Identify these types of boundaries.

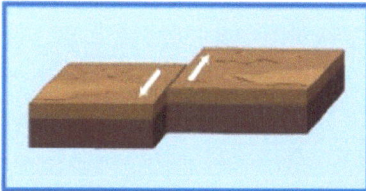

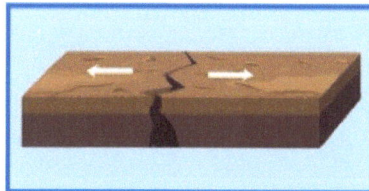

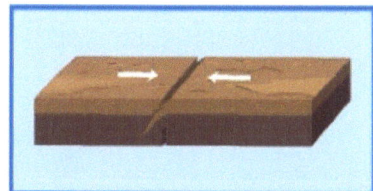

Divergent	Convergent	Transform

Name: _____

Plate Tectonic Quiz

1. What is Pangaea? _____

2. Which of the continents was linked with Europe n Pangaea?
   - a. Africa
   - b. North America
   - c. Asia
   - d. Antarctica

3. The sea surrounding Pangaea is called _____.
   - a. Tethys sea
   - b. Dead sea
   - c. Mediterranean Sea

4. Continental drift occurred due to the movement of _____.
   - a. The oceans
   - b. Tectonic plates
   - c. Drifting plates
   - d. Pangaea plates

5. The Caribbean plate is one of the major plates. True or false?

www.ingramcontent.com/pod-product-compliance
Lightning Source LLC
Chambersburg PA
CBHW052049190326
41521CB00002BA/159